ORIGINS OF BIO-PRODUCTS

The Bio-Industry Backstory

WHAT'S IN THIS BOOK?

PREFACE

In an era marked by environmental consciousness and a growing desire for sustainable living, the concept of bio-products has gained significant prominence. But how did we arrive at this point? What are the stories behind the products we use every day that are derived from nature itself? How have visionaries, scientists, and thought leaders contributed to this thriving industry? These are some of the questions we will explore in the pages ahead.

What the Reader Can Expect

Within these chapters, you will find a comprehensive exploration of the origins and development of bio-products. We will delve into the historical context, trace the emergence of the idea to utilize natural resources, and examine the critical role of technological innovations. We will introduce you to the key figures who have shaped this industry and discuss the challenges they faced along the way.

Moreover, we will address the ever-growing importance of bio-products in addressing environmental concerns and explore their economic and ethical implications on a global scale. You will also discover the trends and potential future innovations that could revolutionize various industries.

As we progress, you will encounter case studies that offer detailed insights into successful bio-products, providing valuable lessons for both consumers and innovators. We will conclude by highlighting your role in the bio-product revolution and offering resources to help you become an active participant in this exciting journey.

So, dear reader, fasten your seatbelts as we embark on a captivating exploration of the past, present, and future of bio-products. Prepare to be inspired by the stories of resilience, creativity, and a shared commitment to a more sustainable and harmonious world.

CHAPTER ONE

UNDERSTANDING BIO-PRODUCTS

In the ever-evolving landscape of modern industry and commerce, the term "bio-products" has become increasingly prevalent. It is a term that encapsulates a vast array of products, each bearing the distinctive mark of their natural origins. In this chapter, we will delve deep into the world of bio-products, exploring their definition and providing an overview of their significance in contemporary society.

Definition of Bio-Products

At its core, a bio-product is a product derived from biological materials or processes. These materials can originate from a variety of sources, including plants, animals, microorganisms, and even waste products. What sets bio-products apart is their reliance on the natural world for their creation, as opposed to traditional synthetic or chemical-based production methods.

Bio-products encompass a wide spectrum of items, ranging from the food we consume to the fabrics we wear, and from the fuels that power our vehicles to the pharmaceuticals that keep us healthy. What unites these diverse products is their shared foundation in biology and the sustainable, eco-friendly approach to their creation.

The Importance in Contemporary Society

In today's world, the significance of bio-products cannot be overstated. As concerns about environmental sustainability, resource depletion, and climate change continue to mount, the bio-products industry has emerged as a beacon of hope and innovation. Here are some key reasons why bio-products are of paramount importance in contemporary society:

Environmental Sustainability: Bio-products are often touted for their reduced environmental impact. Unlike their synthetic counterparts, which may rely on non-renewable resources and generate harmful byproducts, bio-products are typically more sustainable and have a lower carbon footprint.

Reducing Dependency on Fossil Fuels: The bio-products industry plays a pivotal role in the quest for alternative energy sources. Biofuels, for example, are derived from organic materials such as corn or algae and offer a cleaner and more sustainable energy option, reducing our reliance on fossil fuels.

Health and Well-being: Many bio-products are designed with health in mind. Organic foods and natural remedies are sought after by consumers concerned about the potential health risks associated with synthetic additives and chemicals.

Biodegradability: Bio-products are often biodegradable, meaning they can decompose naturally without leaving harmful residues. This quality is particularly important as we grapple with the global issue of plastic pollution.

Biodiversity Preservation: The bio-products industry can have a positive impact on biodiversity by promoting the sustainable cultivation of plants and the conservation of natural habitats.

Economic Opportunities: As the demand for bio-products grows, it opens doors to new economic opportunities, from farming and biotechnology to sustainable manufacturing processes.

Consumer Choice: Increasingly, consumers are seeking products that align with their values, including environmental responsibility and ethical sourcing. Bio-products cater to this growing demand for sustainable and ethically produced goods.

In essence, bio-products represent a harmonious convergence of human innovation and the wisdom of the natural world. They offer solutions to some of the most pressing challenges of our time while respecting the delicate balance of ecosystems.

As we journey through this book, we will uncover the captivating stories and intricate processes that have shaped the world of bio-products. From the earliest historical roots to the cutting-edge technologies of today, we will explore how these products have transformed industries, improved lives, and paved the way for a more sustainable and interconnected future.

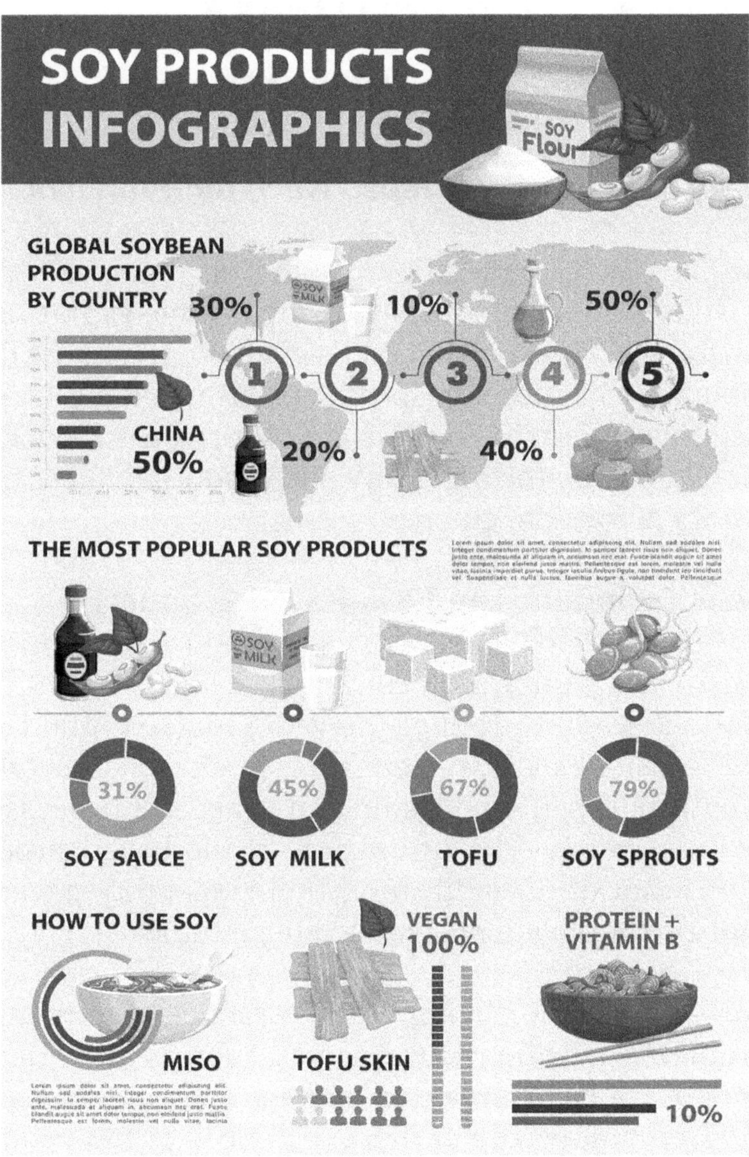

CHAPTER TWO

HISTORICAL CONTEXT

The Global Historical Perspective of Bio-Products

In the pursuit of understanding the origins of bio-products, we embark on a journey through time, exploring the rich historical tapestry that has woven them into the fabric of human civilization. Bio-products, although a buzzword of the modern era, have a lineage that stretches back millennia, shaped by the ingenuity and resourcefulness of our ancestors across the globe.

Ancient Techniques and Bio-Based Products

Our historical exploration commences with ancient civilizations, where bio-products played a fundamental role in daily life. These early societies, often deeply intertwined with nature, relied on organic materials and processes to meet their needs. The utilization of bio-based products was born out of necessity, and their ingenuity paved the way for the bio-products we recognize today.

In ancient Egypt, for instance, the practice of mummification relied heavily on natural resins, such as those derived from pine trees, to preserve bodies. These resins can be considered among the earliest instances of bio-products, showcasing the resourcefulness of ancient cultures.

Similarly, the Silk Road, which connected the East and West, facilitated the exchange of a myriad of bio-products, including spices, dyes, textiles, and medicinal herbs. This historic trade route not only influenced cultures but also allowed for the exchange of valuable biological resources.

Indigenous communities worldwide have cultivated bio-products from the land, harnessing the power of nature to create essential tools, clothing, and medicines. The knowledge passed down through generations, and the sustainable practices developed over centuries, are a testament to the enduring value of bio-products.

The Global Historical Perspective of Bio-Products

As we move through history, we encounter diverse civilizations that have left their indelible mark on the development of bio-products:

•In Asia, the Chinese were pioneers in the use of natural products for medicinal purposes, exploring the healing properties of herbs and plant-based remedies that continue to influence modern medicine.

•Ancient Greece and Rome contributed to the world of cosmetics and perfumes, often using botanical ingredients that laid the foundation for the modern beauty industry.

•Indigenous tribes in the Americas were experts in extracting natural resources for construction, clothing, and sustenance. Their practices highlight the resourcefulness and adaptability of bio-product utilization.

•In Africa, traditional practices included crafting bio-based tools and structures, utilizing materials like bamboo, palm leaves, and animal hides.

•European alchemists sought the secrets of transformation in nature, delving into the extraction and distillation of plant essences to create elixirs and potions.

These historical snapshots demonstrate that bio-products have been intertwined with human civilization since time immemorial. The knowledge, techniques, and wisdom passed down through generations have left an enduring legacy that continues to shape the bio-products industry of today.

Ancient Techniques and Bio-Based Products

To truly understand the roots of bio-products and their enduring relevance in today's world, we must turn the pages of history to explore the ancient techniques and bio-based products that have left an indelible mark on human civilization. These early innovations, born out of necessity and resourcefulness, laid the foundation for the diverse and eco-friendly products we know today.

The Ingenious Beginnings

Ancient civilizations across the globe demonstrated an innate ability to harness the power of nature for their everyday needs. In a world where synthetic alternatives were nonexistent, these early societies relied heavily on organic materials and processes, showcasing remarkable ingenuity.

Preservation and Rituals in Ancient Egypt

One of the most iconic uses of bio-products in ancient times was in the realm of mummification in Egypt. The preservation of bodies for the afterlife required the use of natural resins, primarily derived from trees like pine and cedar. These resins not only helped preserve the bodies but also played a role in the elaborate rituals and beliefs of the time.

The Silk Road and Global Exchange

The historic Silk Road, a vast network of trade routes connecting the East and West, was instrumental in the exchange of bio-based products. This legendary route facilitated the movement of a wide array of goods, including spices, dyes, textiles, and medicinal herbs. The spices of the East, such as cinnamon and pepper, became coveted commodities in the West, while silk, a product of the silkworm, traveled from China to Europe, forever altering cultures and economies along the way.

Indigenous Wisdom and Sustainable Practices

Indigenous communities worldwide epitomized sustainable living long before it became a contemporary trend. They cultivated bio-based products from their local environments, demonstrating a profound connection to the land and a deep understanding of its resources.

Indigenous tribes in North America, for example, crafted functional tools, clothing, and shelters using materials sourced from their surroundings. Birch bark, a flexible and waterproof material, was ingeniously used to construct canoes and shelters. Animal hides were transformed into clothing and footwear, showcasing a profound respect for nature's gifts.

In the Amazon rainforest, indigenous communities have preserved the delicate balance of ecosystems by sustainably harvesting medicinal plants, resins, and other bio-based products. These practices have not only provided for their immediate needs but have also contributed to the global pool of botanical knowledge.

The Lessons of History

These historical snapshots underscore the remarkable adaptability and resourcefulness of ancient civilizations when it came to bio-based products. From preserving the deceased to trading exotic goods and sustaining their communities, our ancestors relied on nature's bounty and harnessed its potential in ways that continue to inspire us.

CHAPTER THREE

EMERGENCE OF THE IDEA

The Shift from Synthetic to Organic: Catalysts

In the journey of understanding the origins of bio-products, we find ourselves at a crucial juncture in history—a moment of transition when humanity began to shift its focus from synthetic materials to organic alternatives. This pivotal transformation was driven by catalysts—innovations, thinkers, and societal changes that reshaped our perception of what was possible and sustainable.

The Dawn of Organic Thinking

As industrialization took hold in the 19th century, synthetic materials and chemical processes rapidly gained prominence. The Industrial Revolution ushered in an era of unprecedented production and innovation, but it also brought with it environmental degradation and health concerns. This dichotomy laid the foundation for a paradigm shift.

Case Studies of the First Known Bio-Products

One of the earliest catalysts for the emergence of bio-products was the realization that organic materials could offer safer and more sustainable alternatives to synthetic counterparts. Two notable case studies from this era exemplify this shift in thinking:

1. Rubber: The Brazilian Connection

In the early 19th century, natural rubber, derived from the latex of rubber trees indigenous to the Amazon rainforest, gained international attention. The water-resistant and elastic properties of rubber made it invaluable for various industrial applications, from waterproof clothing to machinery components. The realization that a natural, renewable resource could serve as a superior alternative to synthetic rubber marked a significant turning point.

2. Dyes: The Age of Natural Colorants

The textile industry was a driving force behind early chemical innovation, particularly in the development of synthetic dyes. However, concerns over the environmental and health impacts of synthetic dyes led to a reevaluation of natural alternatives. Ancient practices of dyeing textiles using plant-based materials were revisited, and the vibrant colors derived from sources like indigo, madder root, and cochineal insects regained popularity. This resurgence of natural dyes marked a shift towards eco-friendly and sustainable textile production.

The Catalysts of Thought

Beyond these specific case studies, the emergence of

bio-products was also shaped by visionary thinkers and movements that challenged the status quo:

1. The Romantic Movement and Nature's Inspiration

The Romantic movement in the 18th and 19th centuries celebrated the beauty and power of nature. Poets, philosophers, and artists began to extol the virtues of the natural world, fostering a profound connection between humanity and the environment. This reverence for nature contributed to a growing desire to find harmony in the materials and processes used in everyday life.

2. Thoreau and the Transcendentalists

In the mid-19th century, writers like Henry David Thoreau and the Transcendentalists advocated for simple living in harmony with nature. Thoreau's experiment at Walden Pond, where he lived deliberately and closely with the land, highlighted the possibilities of a sustainable and purposeful existence. Their ideas laid the philosophical groundwork for a shift towards organic living and bio-based products.

3. Early Environmental Awareness

The late 19th and early 20th centuries saw the emergence of environmental movements and concerns about the impacts of industrialization on the natural world. Pioneers like John Muir, who advocated for the preservation of wilderness areas, and Theodore Roosevelt, who expanded the national park system, raised awareness about the need to protect the environment and its resources.

Case Studies of the First Known Bio-Products

In the annals of history, certain pivotal case studies emerge as beacons of innovation and heralds of the bio-products movement. These early examples provide insight into the human drive to seek sustainable, organic alternatives to synthetic materials and processes.

Rubber: The Brazilian Connection

Natural rubber, extracted from the latex of Hevea brasiliensis trees indigenous to the Amazon rainforest, stands as one of the earliest bio-products to capture international attention. In the early 19th century, the water-resistant and elastic properties of rubber were discovered, making it invaluable for various industrial applications. Waterproof clothing, machinery components, and countless other products benefited from this newfound material. The realization that a natural, renewable resource could serve as a superior alternative to synthetic rubber marked a profound shift in thinking. This case study demonstrated the potential of bio-based materials and their capacity to outperform their synthetic counterparts.

Dyes: The Age of Natural Colorants

The textile industry, a driving force behind early chemical

innovation, heavily relied on synthetic dyes. However, concerns over the environmental and health impacts of these synthetic colorants prompted a reevaluation of natural alternatives. Ancient practices of dyeing textiles using plant-based materials were revisited, and the vibrant colors derived from sources like indigo, madder root, and cochineal insects regained popularity. This resurgence of natural dyes marked a significant shift towards eco-friendly and sustainable textile production. It showcased the potential of bio-products to provide not only color but also safety and sustainability to an industry that was in dire need of change.

These two case studies illuminate the transformative power of bio-products in the midst of industrialization. Natural rubber and plant-based dyes exemplify the capacity of organic materials to offer superior alternatives to synthetic counterparts, while simultaneously addressing environmental and health concerns. They serve as early reminders that sustainable solutions often lie within the embrace of nature itself.

These historical examples are just the beginning of our exploration into the emergence of the bio-products movement. As we venture further into the annals of history, we will uncover additional stories of innovation and adaptation, delving into the individuals and societal changes that catalyzed the transition from synthetic to organic thinking. Together, these narratives paint a vivid picture of the journey towards a more sustainable and eco-conscious world.

Development of Biotechnology (1950 – 2010)

1953

Francis Crick and James Watson describe the double helical structure of DNA.

1955
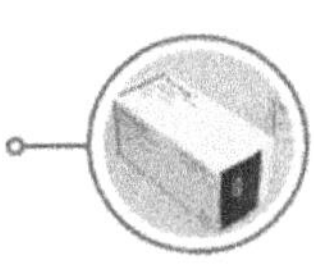
Scientist Frederick Sanger discovered the amino acid sequence of insulin that earned him his first Nobel prize in Chemistry in 1958.

1966

Three scientists shared the 1968 Nobel Prize in Physiology or Medicine for cracking the genetic code for DNA.

1986

First recombinant vaccine, hepatitis B, is approved for human use. Also, the first anticancer drug, interferon, is produced.

1987
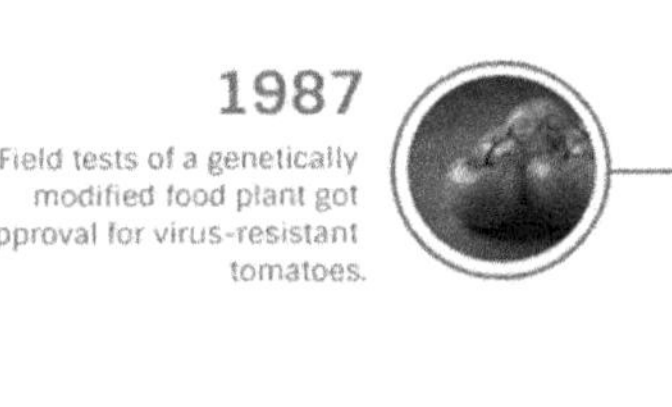
Field tests of a genetically modified food plant got approval for virus-resistant tomatoes.

2001
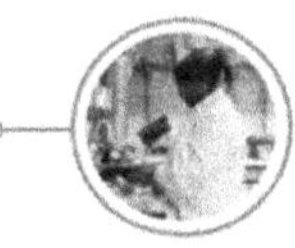
The sequence of the human genome published in Science and Nature. This sequence makes it easy for researchers to start developing treatments.

2004

Scientists at Texas A&M University delivered the world's first cloned pet, a kitten, named CopyCat.

2006
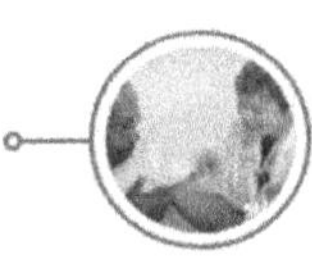
FDA approves a recombinant vaccine against human papillomavirus (HPV) that causes genital warts and can cause cervical cancer.

2010
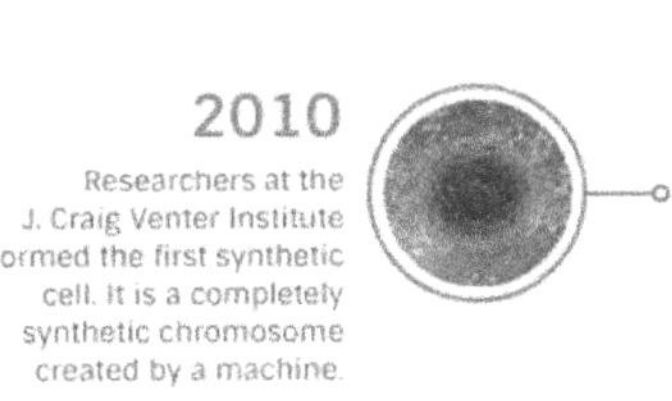
Researchers at the J. Craig Venter Institute formed the first synthetic cell. It is a completely synthetic chromosome created by a machine.

CHAPTER FOUR

ENVIRONMENTAL AWAKENING

Bio-Products and the Environmental Movement

As the 20th century unfolded, a growing sense of environmental consciousness swept across the globe. This era saw the emergence of the environmental movement, a powerful force advocating for the protection of our planet and the promotion of sustainable practices. Within this movement, bio-products found a significant role to play, serving as exemplars of eco-friendly alternatives and contributing to a burgeoning awareness of the interconnectedness between human activity and the environment.

A Shifting Paradigm

The mid-20th century marked a pivotal turning point in human perception. Rapid industrialization and unchecked resource exploitation had left their indelible mark on the environment, leading to increasing pollution, habitat destruction, and concerns about long-term sustainability. In response to these challenges, the environmental movement began to take shape.

Bio-Products as Sustainable Solutions

Bio-products, with their inherent connection to nature and sustainable production methods, quickly became emblematic of the environmental movement's ethos. Here are key aspects of how bio-products contributed to this awakening:

1. Reduced Environmental Impact: Unlike many synthetic counterparts, bio-products were often associated with lower environmental footprints. Whether through biodegradability, reduced chemical usage, or renewable sourcing, bio-products aligned with the goal of minimizing harm to ecosystems.

2. Conservation of Natural Resources: The use of bio-based materials often relied on the cultivation of renewable resources. This sustainable approach contributed to the preservation of ecosystems and the conservation of non-renewable resources.

3. Lower Carbon Footprint: Bio-products, particularly those derived from plants, played a role in reducing greenhouse gas emissions. This was particularly true for biofuels, which offered a cleaner and more sustainable energy source compared to fossil fuels.

4. Alternative to Hazardous Chemicals: In various industries, bio-products provided safer alternatives to toxic chemicals. This shift towards natural, less harmful substances helped protect both human health and the environment.

5. Awareness and Education: The popularity of bio-products helped raise awareness about sustainable living and the importance of making environmentally conscious choices. Consumers increasingly sought out products with eco-friendly credentials.

Influence of Global Environmental Policies

The environmental movement's influence extended beyond grassroots activism. It played a pivotal role in shaping global environmental policies and agreements. Key milestones, such as the establishment of the U.S. Environmental Protection Agency (EPA) in 1970 and the United Nations Conference on the Human Environment in 1972, highlighted the growing importance of environmental issues on the international stage.

Bio-products, with their eco-friendly attributes, were often cited as examples of solutions aligned with these policy goals. The movement also led to regulations and standards that encouraged the development and adoption of bio-products across various industries.

A Symbiotic Relationship

The rise of bio-products and the environmental movement represented a symbiotic relationship. Bio-products provided tangible examples of sustainable alternatives that aligned with the movement's objectives, while the movement, in turn, fostered an environment conducive to the development and adoption of bio-products.

Influence of Global Environmental Policies

The environmental awakening of the 20th century was not confined to grassroots activism and public awareness alone; it also reverberated through the corridors of power and policy on a global scale. This chapter delves into the profound impact of environmental policies and agreements that emerged during this period, and how they intersected with the burgeoning bio-products movement.

The 1970s witnessed the rise of a new consciousness regarding environmental issues. The United States, in a bold step towards addressing environmental challenges, established the Environmental Protection Agency (EPA) in 1970. This milestone marked the beginning of a formalized commitment to safeguarding the environment at the federal level. The EPA became a pivotal force in regulating pollutants, setting standards, and promoting environmental stewardship.

On the international stage, the United Nations Conference on the Human Environment held in Stockholm in 1972 was a watershed moment. The conference brought together nations from around the world to discuss environmental issues and set the stage for global cooperation on environmental matters. The resulting Stockholm Declaration acknowledged the interconnectedness of human activities and the environment, laying the foundation for future international environmental agreements.

One of the most iconic outcomes of this conference was the

establishment of World Environment Day on June 5th, an annual event that continues to serve as a global platform for environmental awareness and action.

The momentum generated by these early initiatives culminated in the 1987 release of the Brundtland Report, officially titled "Our Common Future." Prepared by the World Commission on Environment and Development, this report introduced the concept of sustainable development to the world. It defined sustainable development as "development that meets the needs of the present without compromising the ability of future generations to meet their own needs."

This groundbreaking definition resonated with the global community and solidified sustainability as a central tenet of environmental policy. It recognized that economic, social, and environmental concerns were interconnected and should be addressed holistically.

The 1990s witnessed the signing of several landmark international agreements. The Rio Earth Summit, officially known as the United Nations Conference on Environment and Development (UNCED), held in Rio de Janeiro in 1992, was a pivotal event. The Rio Declaration on Environment and Development, Agenda 21, and the Convention on Biological Diversity were among the key outcomes of this summit.

These agreements emphasized the need for sustainable development, conservation of biodiversity, and the responsible management of natural resources. They elevated environmental issues to the forefront of global priorities and paved the way for a new era of international cooperation.

The Kyoto Protocol of 1997 was another milestone, focusing on reducing greenhouse gas emissions to combat climate change. While not exclusively about bio-products, it underscored the importance of sustainable practices and alternative energy sources, aligning with the broader ethos of the environmental awakening.

The influence of these global environmental policies was twofold. First, they provided a framework for nations to address environmental challenges collectively, setting standards and goals that encouraged sustainability and responsible resource management. Second, they created an environment conducive to the development and adoption of bio-products by emphasizing the importance of eco-friendly alternatives and responsible consumption.

GROUNDWATER
GDEs
SPRINGS
HEADWATERS
CRENAL
KRYAL
STREAMS
LARGE RIVERS
ANCIENT LAKES
LARGE LAKES
A
B
C
D
E
F
G
H
I
J
K
L

CHAPTER FIVE

TECHNOLOGICAL INNOVATIONS SPURRING BIO-PRODUCTS

Role of Technology in Refining Bio-Product Ideas

While the concept of bio-products has ancient roots, it was the convergence of technology and innovation that propelled them into the forefront of contemporary industry and commerce. In this chapter, we explore the pivotal role of technology in refining bio-product ideas and driving the bio-product revolution.

The Modern Biotechnology Revolution

The late 20th century witnessed a revolution in biotechnology, unlocking the full potential of bio-products. Biotechnology, often referred to as "biotech," encompasses a range of techniques and processes that harness living organisms, cells, and molecules to develop new products and technologies. This field of science transformed the way we think about bio-products and their applications.

Genetic Engineering: One of the most transformative developments in biotechnology was the advent of genetic engineering. This technology allowed scientists to manipulate the genetic material of organisms, enabling the creation of bio-products with specific traits or capabilities.

For example, genetically modified crops were engineered to resist pests or tolerate harsh environmental conditions, enhancing crop yields and reducing the need for synthetic pesticides and herbicides.

Fermentation: Fermentation, a process that had been utilized for centuries in food and beverage production, found new applications in bio-products. Microorganisms like bacteria and yeast were harnessed to produce bio-based materials such as enzymes, pharmaceuticals, and biofuels. Fermentation not only provided an eco-friendly alternative to chemical synthesis but also allowed for the scalable production of bio-products.

Enzyme Engineering: Enzymes, natural catalysts that drive chemical reactions in living organisms, became indispensable in the development of bio-products. Advances in enzyme engineering enabled the customization of enzymes to perform specific functions in industrial processes. This innovation reduced energy consumption, waste production, and the reliance on harsh chemicals in manufacturing.

The Emergence of Bioplastics

Bioplastics, a subset of bio-products, exemplify the transformative power of technology in reimagining traditional materials. These bio-based plastics are derived from renewable sources such as corn starch, sugarcane, or potato starch, as opposed to petroleum-based plastics. Bioplastics offer a sustainable alternative with a reduced environmental footprint.

Polylactic Acid (PLA): PLA is a bioplastic derived from corn starch and has gained prominence in packaging and disposable products. Its biodegradability and reduced carbon emissions during production make it an environmentally friendly choice.

Polyhydroxyalkanoates (PHAs): PHAs are a group of biodegradable bioplastics produced by microorganisms through fermentation. They have applications in packaging, medical devices, and agricultural films, offering a sustainable alternative to conventional plastics.

Polyethylene Alternatives: Researchers have also explored bio-based alternatives to traditional polyethylene, one of the most widely used plastics. Bio-based polyethylene, made from sugarcane ethanol, exhibits similar properties to its petroleum-based counterpart but with a smaller carbon footprint.

Bio-Based Chemicals and Pharmaceuticals

Technology has played a significant role in the development of bio-based chemicals and pharmaceuticals, revolutionizing the production of these critical bio-products.

Bio-Based Surfactants: Surfactants, used in a wide range of products from detergents to personal care items, are traditionally derived from petrochemicals. However, bio-based surfactants produced through biotechnology offer a sustainable alternative. They exhibit comparable performance while reducing the environmental impact.

Biopharmaceuticals: Biopharmaceuticals, including vaccines and therapeutic proteins, are increasingly produced using biotechnology techniques. This allows for precise control over the production process, leading to safer and more effective bio-products.

The Future of Bio-Products and Technology

As technology continues to advance, the potential for bio-products to transform industries and reduce their environmental impact grows exponentially. Innovations in synthetic biology, nanotechnology, and materials science promise to expand the scope and capabilities of bio-products further.

Biotechnology and Its Impact

The bio-product revolution owes much of its success to the transformative influence of biotechnology. Biotechnology, often dubbed "biotech," represents a multidisciplinary field that harnesses the power of living organisms, cells, and molecules to develop innovative products and technologies. In this chapter, we explore the profound impact of biotechnology on the development and proliferation of bio-products.

Unlocking the Potential of Genetic Engineering

Genetic engineering emerged as a game-changer within the biotechnology landscape, enabling scientists to manipulate the genetic material of organisms for specific outcomes. This breakthrough had far-reaching implications for the creation and refinement of bio-products.

Genetically Modified Organisms (GMOs): GMOs are organisms whose genetic material has been altered to introduce desirable traits or characteristics. In agriculture, GMOs have been engineered to resist pests, tolerate adverse environmental conditions, and enhance crop yields. For example, genetically modified crops like Bt cotton produce their insecticides, reducing the need for chemical pesticides and benefiting both farmers and the environment.

Customized Microorganisms: Biotechnology allows researchers to customize microorganisms like bacteria and yeast for industrial applications. These engineered microorganisms serve as bio-factories for producing bio-based materials, pharmaceuticals, and biofuels. The ability to fine-tune microbial processes has resulted in more efficient and sustainable production methods.

Fermentation as a Sustainable Tool

Fermentation, a natural process harnessed by biotechnology, played a pivotal role in refining bio-product ideas. Microorganisms like yeast and bacteria are employed to convert organic materials into valuable bio-products through fermentation processes.

Ethanol Production: Fermentation is used to convert plant-based materials, such as corn or sugarcane, into bio-fuels like ethanol. Ethanol, a bio-based alternative to gasoline, has been widely adopted for its lower environmental impact and potential to reduce greenhouse gas emissions.

Pharmaceuticals and Enzymes: Biotechnology-driven fermentation is also employed in the production of pharmaceuticals and enzymes. Bio-based pharmaceuticals offer precise treatments with fewer side effects, while enzymes serve as catalysts in a variety of industrial processes, reducing the need for harsh chemicals and energy-intensive reactions.

Customized Enzymes and Reduced Environmental Impact

Enzymes, natural catalysts found in living organisms, have been at the forefront of biotechnology-driven innovation. Scientists have harnessed the power of enzymes to create customized biocatalysts, reducing the environmental footprint of various industrial processes.

Enzyme Engineering: By modifying and optimizing enzymes, researchers have developed biocatalysts tailored for specific applications. Enzymes are used in a wide range of industries, including agriculture, food production, and textile manufacturing, to replace traditional chemical processes with more eco-friendly alternatives.

Green Chemistry: The principles of green chemistry, which promote the use of benign and sustainable processes, have been advanced through biotechnology. Bio-based solvents and enzymes have enabled the development of greener chemical reactions, reducing waste and environmental harm.

Bioplastics: A Sustainable Alternative

Bioplastics, derived from renewable sources like corn starch or sugarcane, represent a sustainable alternative to traditional petroleum-based plastics. Biotechnology has played a significant role in their development and proliferation.

Polylactic Acid (PLA): PLA is a bioplastic produced through fermentation of corn starch. It is used in a variety of applications, including packaging materials and disposable items. PLA biodegrades under certain conditions, reducing the persistence of plastic waste.

Polyhydroxyalkanoates (PHAs): PHAs are a class of bioplastics produced by microorganisms through fermentation. They are biodegradable and exhibit properties similar to conventional plastics, making them a sustainable choice for various applications.

Future Bioplastic Innovations: Ongoing biotechnological research holds the promise of creating even more advanced bioplastics with improved properties, potentially replacing a broader range of petroleum-based plastics.

Conclusion

Biotechnology has revolutionized the development and impact of bio-products across multiple industries. From genetically modified crops and customized microorganisms to sustainable bioplastics and greener industrial processes, biotechnology has expanded the possibilities for bio-products while reducing their environmental footprint.

CHAPTER SIX

KEY FIGURES AND THOUGHT LEADERS

Profiles of Individuals Pivotal in Bio-Product Conceptualization

The journey of bio-products from conception to revolution has been steered by the visionary thinkers, researchers, and entrepreneurs who dared to challenge the status quo. In this chapter, we explore the profiles of key figures and thought leaders whose contributions have played a pivotal role in shaping the bio-product landscape.

Rachel Carson (1907-1964): The Environmental Visionary

Rachel Carson, a marine biologist and author, stands as a towering figure in the environmental movement and the conceptualization of eco-friendly bio-products. Her groundbreaking book, "Silent Spring" (1962), exposed the detrimental effects of synthetic pesticides, particularly DDT, on the environment and wildlife. Carson's powerful prose and fearless advocacy sparked public awareness and led to the eventual ban on DDT in the United States. Her work set a precedent for the environmental consciousness that would later drive the demand for bio-products aligned with nature.

Aldo Leopold (1887-1948): The Ethical Steward

Aldo Leopold, often referred to as the father of modern ecology, was a forester, ecologist, and environmentalist who emphasized the importance of ethical land stewardship. His book, "A Sand County Almanac" (1949), introduced the concept of a "land ethic," advocating for the responsible treatment of the land and its resources. Leopold's ideas laid the ethical foundation for the bio-products movement, highlighting the interconnectedness of nature and human responsibility.

Rudolf Steiner (1861-1925): The Pioneer of Biodynamics

Rudolf Steiner, an Austrian philosopher and founder of biodynamic agriculture, introduced holistic farming practices that emphasized the interconnectedness of soil, plants, animals, and humans. Biodynamic agriculture, a precursor to modern organic farming, promotes sustainable and harmonious farming methods. Steiner's vision of agriculture that respects nature and relies on bio-based solutions continues to influence the production of bio-products rooted in organic principles.

Norman Borlaug (1914-2009): The Agricultural Innovator

Norman Borlaug, an American agronomist and humanitarian, is credited with spearheading the Green Revolution. His work in developing high-yield, disease-resistant wheat varieties contributed to a dramatic increase in global food production, saving millions from hunger and malnutrition. Borlaug's bio-product innovations not only revolutionized agriculture but also highlighted the potential of bio-based solutions to address critical global challenges.

Barbara McClintock (1902-1992): The Genetic Pioneer

Barbara McClintock, an American geneticist and Nobel laureate, made groundbreaking contributions to the field of genetics. Her discovery of transposons, or "jumping genes," revolutionized our understanding of genetic variability and adaptation in organisms. McClintock's work in deciphering the genetic code paved the way for genetic engineering and the development of bio-products with customized traits.

Their Contributions and Influence

The bio-product revolution owes much of its success to the contributions and influence of key figures and thought leaders who championed the cause of sustainable and eco-friendly innovations. In this chapter, we delve deeper into their individual impacts on the bio-product landscape and the lasting legacy they have left behind.

Rachel Carson (1907-1964): The Environmental Visionary

Contributions: Rachel Carson's most influential work, "Silent Spring" (1962), exposed the detrimental effects of synthetic pesticides on the environment, wildlife, and human health. Through meticulous research and compelling prose, Carson revealed the interconnectedness of all living beings and ecosystems, emphasizing the fragility of our planet.

Influence: Carson's work ignited a global environmental movement, inspiring a wave of activism and policy changes. The ban on DDT in the United States in 1972 was a direct result of her advocacy. Her legacy lives on in the countless individuals and organizations dedicated to environmental conservation and the pursuit of bio-products aligned with nature.

Aldo Leopold (1887-1948): The Ethical Steward

Contributions: Aldo Leopold's book, "A Sand County Almanac" (1949), introduced the concept of a "land ethic," emphasizing the ethical responsibility of humans to treat the land and its resources with care and respect. He advocated for responsible land management and conservation practices, including those that underpin organic farming and bio-product production.

Influence: Leopold's land ethic has had a profound influence on environmental ethics and policy, shaping the way we perceive our relationship with the natural world. His ideas continue to guide conservation efforts and the development of sustainable bio-products, reinforcing the connection between ethics and eco-conscious choices.

Rudolf Steiner (1861-1925): The Pioneer of Biodynamics

Contributions: Rudolf Steiner's pioneering work in biodynamic agriculture introduced holistic and sustainable farming practices. He emphasized the importance of viewing the farm as a self-sustaining organism, using bio-based solutions to enhance soil health, crop quality, and biodiversity. Steiner's lectures on biodynamics laid the foundation for modern organic farming.

Influence: Steiner's biodynamic principles continue to inspire organic farming practices worldwide. Biodynamic agriculture emphasizes a balanced and holistic approach to farming, aligning with the principles of bio-product production. Steiner's legacy lives on in the organic and sustainable farming movements.

Norman Borlaug (1914-2009): The Agricultural Innovator

Contributions: Norman Borlaug's groundbreaking work in developing high-yield, disease-resistant wheat varieties during the Green Revolution transformed global agriculture. His innovations significantly increased food production and played a pivotal role in alleviating global hunger and malnutrition.

Influence: Borlaug's contributions to agriculture and bio-product innovations earned him the Nobel Peace Prize in 1970. His legacy continues to inspire advancements in sustainable agriculture and bio-based solutions to address food security challenges worldwide.

Barbara McClintock (1902-1992): The Genetic Pioneer

Contributions: Barbara McClintock's discovery of transposons, or "jumping genes," revolutionized the field of genetics. Her work revealed the dynamic nature of genetic information and laid the foundation for genetic engineering and the development of bio-products with customized traits.

Influence: McClintock's pioneering research earned her the Nobel Prize in Physiology or Medicine in 1983. Her contributions continue to shape the field of biotechnology, influencing the development of bio-products, genetically modified organisms, and personalized medicine.

Conclusion:

These key figures and thought leaders have left an enduring legacy that extends beyond their lifetimes. Their contributions and influence have driven the bio-product revolution, emphasizing the importance of ethical stewardship, environmental responsibility, and innovative solutions. As we navigate the evolving landscape of bio-products, we draw inspiration from their visionary ideas and continue to build upon their remarkable contributions. In the chapters that follow, we will explore the challenges faced in the early stages of the bio-product revolution and examine the evolution of consumer consciousness toward sustainable and eco-friendly choices, further illuminating the path forward in the bio-product movement.

The rise of biologicals, among other tools and practices to drive sustainability, is part of a greater shift in the duty of care of large Ag corporations to people and planet.

Paula Pinto

CHAPTER SEVEN

CHALLENGES IN THE EARLY STAGES

Societal, Financial, and Technical Hurdles

The path to the bio-product revolution was not without its share of challenges and obstacles. In this chapter, we explore the societal, financial, and technical hurdles faced during the early stages of the bio-product movement and the innovative solutions that emerged in response.

Societal Hurdles

Lack of Awareness: In the early days of the bio-product revolution, many consumers and businesses were unaware of the environmental and sustainability issues associated with conventional products. Raising awareness about the benefits of bio-products and the environmental impact of alternatives was a significant challenge.

Consumer Skepticism: Skepticism about the efficacy and affordability of bio-products was prevalent. Consumers questioned whether these products could perform as well as their synthetic counterparts. Overcoming this skepticism required transparent communication and product demonstrations.

Market Resistance: Established industries often resisted the adoption of bio-products, viewing them as disruptive innovations that threatened existing business models. This resistance presented hurdles for bio-product entrepreneurs seeking market acceptance.

Financial Hurdles

High Initial Costs: Developing and producing bio-products, especially those derived from novel biotechnological processes, often incurred higher initial costs compared to traditional alternatives. Securing funding for research, development, and production posed a challenge.

Market Entry Barriers: Entering competitive markets dominated by conventional products required substantial financial investments in marketing, distribution, and supply chain infrastructure. Many bio-product startups faced barriers to entry due to limited financial resources.

Price Competitiveness: Bio-products initially struggled to compete on price, as economies of scale favored traditional production methods. Lowering production costs without compromising quality was a financial challenge for bio-product manufacturers.

Technical Hurdles

Biotechnology Complexity: The integration of biotechnology into bio-product development was technically demanding. Manipulating genes, optimizing microbial processes, and producing bio-based materials at scale required specialized expertise and research.

Regulatory Uncertainty: Regulatory frameworks for bio-products were often ambiguous or underdeveloped in the early stages of the movement. Bio-product manufacturers faced uncertainty about compliance and product safety, which hindered market entry.

Scaling Up Production: Transitioning from small-scale laboratory production to large-scale manufacturing presented technical challenges. Maintaining product consistency and quality while increasing production capacity required innovative solutions.

Innovative Responses to Hurdles

Education and Outreach: Overcoming societal hurdles involved extensive education and outreach efforts. Environmental organizations, government agencies, and industry associations played a crucial role in disseminating information about the benefits of bio-products.

Transparency and Certification: Building trust with consumers required transparency in product labeling and certification. Eco-labels, such as the USDA Organic seal, provided a means for consumers to identify genuine bio-products.

Research and Development Funding: Governments and private investors began to recognize the potential of bio-products and provided funding for research and development. This financial support enabled bio-product start-ups to innovate and refine their offerings.

Collaborative Research: Collaborative research initiatives brought together scientists, entrepreneurs, and industry stakeholders to address technical challenges. These partnerships accelerated advancements in biotechnology and process optimization.

Regulatory Frameworks: Governments developed clearer regulatory frameworks for bio-products, ensuring product safety and standardization. This regulatory clarity reduced uncertainty for manufacturers and facilitated market entry.

Technological Advancements: Technical hurdles were met with continuous advancements in biotechnology and materials science. Breakthroughs in enzyme engineering, fermentation, and bioprocessing improved production efficiency and product quality.

Overcoming Misconceptions and Technical Limitations

The early stages of the bio-product revolution were characterized by a battle on two fronts: dispelling misconceptions about bio-products and addressing the technical limitations that hindered their development and adoption. In this chapter, we explore the challenges faced and the innovative solutions that emerged to overcome these hurdles.

Misconceptions About Bio-Products

Perceived Ineffectiveness: One of the primary misconceptions surrounding bio-products was the belief that they were less effective than their synthetic counterparts. Some consumers and businesses questioned whether bio-based materials could deliver the same level of performance, whether in cleaning products, packaging, or industrial applications.

Affordability Concerns: Another common misconception was that bio-products would be more expensive due to their environmentally friendly and sustainable attributes. This perception often deterred cost-conscious consumers from choosing bio-products, believing that they were a premium or niche offering.

Greenwashing and False Claims: The rise of bio-products brought forth instances of greenwashing—false or misleading claims about a product's environmental benefits. This deceptive marketing further eroded consumer trust and complicated the marketplace.

Overcoming Misconceptions

Transparency and Education: Addressing misconceptions required transparency and education efforts by bio-product manufacturers. Companies invested in clear labeling, certifications, and consumer education campaigns to convey the benefits and efficacy of bio-products accurately.

Product Performance Testing: Demonstrating the effectiveness of bio-products through rigorous testing and performance comparisons with synthetic alternatives helped alleviate concerns about their capabilities. Scientific studies and third-party evaluations provided empirical evidence of their efficacy.

Price Competitiveness: To overcome the affordability misconception, manufacturers focused on achieving price competitiveness through economies of scale, efficient production processes, and sourcing cost-effective bio-based materials.

Technical Limitations

Biotechnological Challenges: Early-stage bio-products faced biotechnological challenges, particularly in the development of genetically modified organisms (GMOs) and customized microbial processes. Manipulating genes and optimizing bioprocesses required specialized expertise and research.

Regulatory Uncertainty: Regulatory frameworks for bio-products were often ambiguous or underdeveloped in the early stages of the movement. Manufacturers faced uncertainty about compliance and product safety, which hindered market entry.

Scale-Up Complexities: Transitioning from laboratory-scale production to large-scale manufacturing posed technical difficulties. Maintaining product consistency, quality, and yield while increasing production capacity was a significant challenge.

Overcoming Technical Limitations

Research and Development: Significant investments in research and development allowed bio-product manufacturers to address biotechnological challenges and refine production processes. Advancements in enzyme engineering, fermentation techniques, and genetic modifications improved efficiency and scalability.

Collaborative Innovation: Collaborative research initiatives brought together scientists, entrepreneurs, and industry stakeholders to pool resources and expertise. These partnerships accelerated advancements in biotechnology and bioprocessing, helping overcome technical limitations.

Regulatory Frameworks: Governments recognized the importance of clear regulatory frameworks for bio-products. The development of comprehensive regulations ensured product safety and standardization, providing manufacturers with the guidance needed for market entry.

CHAPTER EIGHT

EVOLUTION OF CONSUMER CONSCIOUSNESS

The Growing Demand for Sustainable and Organic Products

Consumer consciousness, the collective awareness and values of individuals as they make purchasing decisions, has evolved significantly over the past few decades. A prominent facet of this transformation is the burgeoning demand for sustainable and organic products. In this chapter, we delve into the factors driving this shift and the profound impact it has had on the bio-products industry.

A Paradigm Shift in Consumer Values

The late 20th and early 21st centuries witnessed a notable shift in consumer values and priorities. This transformation can be attributed to several key factors:

1. Environmental Awareness: As the environmental movement gained momentum, consumers became increasingly cognizant of the environmental impact of their choices. Concerns about climate change, pollution, and habitat destruction spurred a desire to make eco-conscious decisions.

2. Health and Well-being: Rising awareness of the potential health risks associated with synthetic additives, chemicals, and genetically modified organisms (GMOs) led consumers to seek healthier, more natural alternatives. Organic foods and products, free from synthetic pesticides and additives, gained popularity.

3. Ethical and Social Considerations: Consumers began to scrutinize the ethical practices of companies, demanding transparency in sourcing, fair labor practices, and humane treatment of animals. This ethical dimension of consumer consciousness extended to bio-products that aligned with these values.

4. Desire for Quality: Many consumers came to associate bio-products with higher quality and authenticity. They appreciated the craftsmanship, attention to detail, and dedication to natural ingredients often found in bio-based offerings.

The Rise of Sustainable and Organic Products

Against this backdrop of shifting consumer values, the demand for sustainable and organic products experienced exponential growth. Here are key manifestations of this trend:

1. Organic Foods: The organic food sector saw remarkable expansion, with consumers seeking organic produce, dairy, and meat products. The USDA Organic label in the United States and equivalent certifications worldwide provided a recognizable standard for organic products.

2. Natural and Organic Personal Care: Cosmetics, skincare, and personal care products underwent a transformation as consumers sought natural alternatives free from synthetic chemicals and harmful ingredients. Natural and organic personal care brands gained prominence.

3. Sustainable Fashion: The fashion industry witnessed a surge in sustainable and eco-friendly clothing brands. Consumers began to prioritize clothing made from organic fibers, such as cotton and hemp, and sought products produced under fair labor conditions.

4. Eco-Friendly Cleaning Products: Consumers shifted towards eco-friendly cleaning solutions, opting for bio-based alternatives that were safer for their homes and the environment.

5. Renewable Energy: The renewable energy sector experienced unprecedented growth as consumers embraced clean energy sources, such as solar and wind power, in their homes and businesses.

Impact on the Bio-Products Industry

The growing demand for sustainable and organic products presented a significant opportunity for the bio-products industry. Bio-based products, with their natural ingredients, renewable sourcing, and eco-friendly production methods, aligned perfectly with the values of environmentally conscious consumers.

This shift in consumer consciousness prompted companies to reevaluate their practices and offerings. Many businesses began incorporating bio-based materials and sustainable processes into their product lines. The bio-products industry, once considered a niche, found itself at the forefront of consumer preferences.

Challenges and Future Trends

While the evolution of consumer consciousness towards sustainability and organic choices has been transformative, it has also posed challenges. The bio-products industry must meet the growing demand while maintaining high standards of quality, transparency, and ethical sourcing.

Looking ahead, the trajectory of consumer consciousness appears to be on a continued path towards sustainability. Consumers are likely to demand even greater transparency, traceability, and environmental accountability from the products they purchase. The bio-products industry, with its foundation in eco-friendly principles, is well-positioned to thrive in this evolving landscape.

Marketing and Consumer Education Strategies

The transformation in consumer consciousness towards sustainability and organic choices has been propelled not only by changing values but also by effective marketing and consumer education strategies. In this chapter, we explore the vital role played by marketing initiatives and educational efforts in driving the demand for sustainable and organic products, particularly within the bio-products industry.

Building Awareness and Trust

One of the foundational pillars of consumer education has been raising awareness about the benefits of sustainable and organic products. Companies within the bio-products sector have invested in educational campaigns that highlight the positive impact of these products on the environment, personal health, and ethical considerations. These campaigns often emphasize the transparency and traceability of sourcing, production methods, and the commitment to eco-friendly practices. Such messaging helps consumers make informed choices and builds trust in the products and brands that align with their values.

Certifications and Labeling

Certifications and labeling have played a crucial role in simplifying the decision-making process for consumers. Recognizable symbols, such as the USDA Organic label, Fair Trade certification, or cruelty-free logos, provide consumers with quick and reliable indicators of a product's environmental and ethical attributes. These certifications serve as a shorthand for the values that conscientious consumers seek, allowing them to make choices that align with their principles.

Transparency in Supply Chains

In an era marked by increased scrutiny of supply chains, the bio-products industry has emphasized transparency. Companies have adopted transparent sourcing practices, sharing information about the origins of raw materials, the conditions under which products are manufactured, and the environmental impact of their production processes. This transparency not only builds consumer trust but also holds companies accountable for their claims regarding sustainability and ethical practices.

Educational Content and Resources

Consumer education extends beyond marketing campaigns to encompass the creation of educational content and resources. Many bio-products brands have developed informative websites, blogs, and social media channels that provide consumers with valuable information about sustainable living, the benefits of organic products, and the environmental impact of their choices. These resources empower consumers to make more informed decisions and encourage them to adopt a sustainable lifestyle.

Collaboration and Advocacy

Some bio-products companies actively engage in collaborations with environmental organizations, NGOs, and advocacy groups. By partnering with entities that share their values, these companies amplify their reach and impact. Collaborative efforts often involve initiatives such as tree planting, conservation projects, and sustainability campaigns, which resonate with consumers and underscore the commitment of brands to their principles.

Consumer Feedback and Engagement

Successful bio-products companies actively engage with their consumers and seek feedback. This two-way communication allows companies to understand consumer preferences, address concerns, and adapt their products and practices accordingly. Engaging with consumers on social media platforms, conducting surveys, and responding to reviews are some of the ways brands foster a sense of community and build lasting relationships.

The Road Ahead

As the bio-products industry continues to grow and consumer consciousness evolves, marketing and education strategies will remain instrumental in shaping the market landscape. Companies that effectively communicate their commitment to sustainability, ethical sourcing, and environmental responsibility are likely to thrive in this increasingly eco-conscious world. Moreover, the bio-products industry's dedication to innovation and transparency will play a pivotal role in educating consumers and driving demand for sustainable and organic products. As we explore this chapter further, we will delve into specific marketing campaigns, success stories, and strategies employed by bio-products brands that have contributed to the expansion of consumer consciousness in favor of sustainable and organic choices.

CHAPTER NINE

ECONOMIC AND ETHICAL IMPLICATIONS

Bio-Products in the Global Market: Economic Impact

The ascent of bio-products within the global market has not only transformed industries but also led to significant economic implications. In this chapter, we delve into the economic impact of bio-products, exploring how they have shaped markets, created new opportunities, and influenced the global economy.

Market Growth and Expansion

One of the most prominent economic impacts of bio-products has been their role in expanding existing markets and creating entirely new ones. As consumer demand for sustainable and eco-friendly alternatives has grown, industries have adapted to meet these needs. Bio-products have driven market growth in sectors such as organic food, renewable energy, and green technology.

Job Creation and Industry Growth

The bio-product revolution has spurred job creation across various sectors. Bio-based industries, from agriculture and biotechnology to manufacturing and renewable energy, have witnessed significant growth. As demand for bio-products continues to rise, so does the need for skilled labor, research and development, and production capacity.

Investment and Innovation

Investors have recognized the economic potential of bio-products and have poured capital into research, development, and production. This investment has led to continuous innovation, driving advancements in biotechnology, sustainable agriculture, and green chemistry. Startups and established companies alike have been able to access funding to develop and scale their bio-product offerings.

Competitive Advantage and Market Differentiation

Companies that have embraced bio-products have gained a competitive edge in the marketplace. Bio-products are often positioned as premium, eco-friendly alternatives to conventional products. Businesses that offer bio-based solutions can differentiate themselves, capture market share, and command higher prices for their environmentally responsible offerings.

Reduced Environmental Costs

Bio-products, by their nature, tend to have a lower environmental impact compared to their synthetic counterparts. This translates into reduced environmental costs for industries and society as a whole. Lower emissions, reduced resource consumption, and decreased pollution contribute to long-term economic benefits by mitigating environmental damage and related expenses.

Challenges and Costs

While the economic benefits of bio-products are significant, challenges and costs must also be considered. The transition to bio-based production methods may require initial capital investments, retooling of manufacturing processes, and adaptation to new supply chains. Companies must navigate regulatory compliance and ensure that their bio-products meet stringent quality standards.

Impact on Traditional Industries

The rise of bio-products has had a disruptive impact on traditional industries, particularly those reliant on fossil fuels and petrochemicals. Biofuels, for example, have challenged the dominance of the oil industry, while bioplastics are gradually reducing the demand for petroleum-based plastics. As the bio-product revolution continues, these traditional industries are faced with the need to adapt and innovate.

Global Trade and Sustainability

Bio-products have also influenced global trade dynamics. As sustainability becomes a driving force in consumer choices, countries that can offer bio-products with robust eco-credentials gain a competitive advantage in international markets. This shift has the potential to rebalance global trade patterns and promote sustainability on a global scale.

Ethical Considerations in Production and Consumption

The rise of bio-products within the global market has not only reshaped economies but has also sparked crucial ethical considerations. In this chapter, we delve into the ethical implications surrounding the production and consumption of bio-products, exploring their role in promoting responsible choices and sustainable practices.

Responsible Sourcing and Production

One of the fundamental ethical considerations in bio-product production lies in responsible sourcing. Sustainable and ethical sourcing of raw materials is paramount to ensuring that bio-products align with eco-friendly principles. This includes practices such as fair labor standards, biodiversity protection, and ethical treatment of animals when relevant.

Reduction of Environmental Impact

Bio-products are often favored for their reduced environmental impact compared to synthetic alternatives. Ethical production of bio-products prioritizes minimizing carbon footprints, resource depletion, and pollution. This commitment to environmental responsibility extends to using renewable energy sources, adopting circular economy practices, and minimizing waste throughout the production process.

Ethical Labor Practices

The bio-product movement places a strong emphasis on ethical labor practices. From farmers and harvesters in agriculture to factory workers in manufacturing, ensuring fair wages, safe working conditions, and equitable treatment are vital ethical considerations. This commitment extends to the entire supply chain, from sourcing to distribution.

Consumer Awareness and Empowerment

Ethical considerations extend to the empowerment of consumers. Bio-product manufacturers often engage in transparent labeling and information sharing, allowing consumers to make informed choices. This transparency enhances consumer empowerment and promotes ethical consumption by enabling individuals to align their values with their purchasing decisions.

Protection of Biodiversity

Bio-products sourced from agriculture or natural habitats may impact biodiversity. Ethical production takes steps to protect and enhance biodiversity by avoiding habitat destruction, supporting sustainable land management practices, and prioritizing organic farming methods that encourage biodiversity.

Reduction of Chemical Usage

Bio-products often boast reduced chemical usage compared to synthetic counterparts. Ethical production minimizes the use of harmful chemicals, pesticides, and fertilizers that can negatively impact ecosystems, human health, and wildlife. This focus on chemical reduction aligns with responsible and sustainable agriculture.

Promotion of Renewable Resources

Ethical considerations emphasize the promotion of renewable resources. Bio-products harness the power of nature and renewable materials. This reduces the dependency on finite resources and encourages a more sustainable approach to production, aligning with the principles of responsible resource management.

Addressing Social and Environmental Justice

Ethical production also addresses issues of social and environmental justice. It considers the impact of production on marginalized communities, indigenous rights, and vulnerable ecosystems. Ethical bio-product manufacturers actively engage in practices that uphold justice and equity.

Balancing Profit and Purpose

Ethical bio-product companies often embrace a "profit with purpose" ethos. While pursuing economic success, they also strive to make a positive impact on society and

the environment. This dual commitment to profit and purpose reflects the ethical dimension of bio-product entre-
preneurship.

CHAPTER TEN

FUTURE OF BIO-PRODUCTS

Current Trends and Future Innovations

The bio-product revolution has laid a foundation for sustainable, eco-friendly alternatives in various industries. In this chapter, we explore the current trends and anticipate future innovations that promise to further shape the landscape of bio-products.

Bioplastics and Packaging

Current Trend: Bioplastics, derived from renewable sources like corn starch or sugarcane, have gained traction as a sustainable alternative to conventional plastics. They are being used in packaging, reducing the environmental impact of plastic waste.

Future Innovation: Innovations in bioplastics are expected to yield materials with enhanced durability and versatility, making them suitable for a wider range of applications, including electronics and automotive industries. Additionally, biodegradable and compostable bioplastics will become more prevalent, reducing the persistence of plastic waste in the environment.

Biofuels and Renewable Energy

Current Trend: Biofuels, such as biodiesel and bioethanol, have gained prominence as cleaner alternatives to fossil fuels. They are used in transportation and energy generation.

Future Innovation: Advancements in biofuel production will focus on improving feedstock efficiency, increasing energy yields, and optimizing the production process. Second-generation biofuels, derived from non-food biomass like agricultural residues, will become more viable, reducing competition with food crops.

Biopharmaceuticals and Medicine

Current Trend: Bio-products in the medical field include biopharmaceuticals like insulin and vaccines, produced using biotechnology.

Future Innovation: Future innovations in biopharmaceuticals will focus on personalized medicine, with treatments tailored to an individual's genetic profile. Advanced bioprocessing techniques will enable cost-effective production of complex biologics, expanding access to cutting-edge medical treatments.

Biological Pest Control

Current Trend: Bio-based alternatives for pest control, such as biopesticides and beneficial insects, are being employed in agriculture to reduce the use of chemical pesticides.

Future Innovation: Continued research into biopesticides and precision agriculture techniques will lead to more effective and targeted pest control methods. The integration of artificial intelligence and sensors will enable real-time monitoring and optimized pest management strategies.

Biotechnology and Genetic Engineering

Current Trend: Genetic engineering plays a crucial role in bio-product development. It has led to the creation of genetically modified crops, enzymes for industrial processes, and bio-based materials with tailored properties.

Future Innovation: Advances in gene editing technologies like CRISPR-Cas9 will revolutionize biotechnology. This will allow for precise genetic modifications in crops for increased yield and resilience, as well as the development of novel bio-based materials with customizable properties.

Circular Economy and Waste Reduction

Current Trend: The circular economy model is gaining traction, emphasizing recycling, reuse, and waste reduction. Bio-products play a vital role in this model, offering biodegradable and compostable materials.

Future Innovation: Bio-products will continue to contribute to a circular economy by promoting the use of renewable resources and reducing waste. Innovations in waste-to-energy technologies and biodegradable materials will drive the transition toward a more sustainable and closed-loop system.

Consumer Demand for Transparency

Current Trend: Consumers increasingly demand transparency and traceability in product supply chains. Ethical and eco-friendly certifications, such as Fair Trade and USDA Organic, are sought after by consumers.

Future Innovation: Blockchain technology and IoT (Internet of Things) devices will provide consumers with real-time information about the origin, production, and environmental impact of products. This increased transparency will empower consumers to make informed choices.

The Potential Impact on Global Industries

The future of bio-products holds the potential to revolutionize global industries across the board. These sustainable and eco-friendly alternatives are poised to make a substantial impact on sectors ranging from agriculture and manufacturing to healthcare and energy. As the bio-product revolution gains momentum, it promises to reshape how industries operate and address some of the world's most pressing challenges.

In agriculture, bio-products are expected to play a pivotal role in sustainable food production. Innovations in biotechnology, genetic engineering, and precision agriculture will lead to increased crop yields, reduced resource consumption, and minimized environmental impact. Bio-based pesticides and fertilizers will replace harmful chemicals, promoting healthier ecosystems and safer food production.

The manufacturing industry is set to undergo a significant transformation driven by bio-products. Bioplastics and bio-based materials will replace traditional plastics and synthetic materials, reducing the carbon footprint and dependence on fossil fuels. 3D printing with bio-based materials will become more widespread, allowing for customizable, eco-friendly manufacturing processes.

The healthcare and pharmaceutical sectors will see continued growth in biopharmaceuticals and personalized medicine. Advanced bioprocessing techniques will enable cost-effective production of complex biologics, providing patients with more effective and tailored treatments. Gene editing technologies like CRISPR-Cas9 will revolutionize genetic therapies and disease prevention.

In the energy sector, biofuels will become an integral part of the transition to renewable energy sources. Second-generation biofuels, derived from non-food biomass, will reduce competition with food crops and provide a sustainable alternative to fossil fuels. Bioenergy will play a critical role in decentralized and clean energy production.

The waste management and recycling industry will benefit from the growth of bio-products. Biodegradable and compostable materials will contribute to a circular economy by reducing waste and promoting sustainable consumption. Innovations in waste-to-energy technologies will turn organic waste into renewable energy sources.

Global trade patterns will shift as countries that excel in bio-product production gain a competitive advantage. Sustainable and eco-friendly products will become a focal point in international trade, influencing supply chains and trade agreements. Developing nations with abundant natural resources may see economic growth through the sustainable production of bio-based materials.

The potential impact of bio-products on global industries extends beyond economic considerations. Ethical and environmental principles will guide industries toward responsible and sustainable practices. Transparency in supply chains will empower consumers to make informed choices, fostering a sense of social and environmental responsibility.

As industries adapt to the bio-product revolution, collaboration between governments, businesses, and research institutions will be crucial. Regulatory frameworks will need to evolve to support the growth of bio-products while ensuring safety and sustainability. Investments in research, development, and infrastructure will drive innovation and competitiveness.

In conclusion, the future of bio-products holds ir
promise for transforming global industries. Their
able and eco-friendly nature aligns with the ;
demand for responsible and ethical alternatives. A
tries embrace bio-products, they are poised to
pressing environmental challenges while driving e
ic growth and innovation on a global scale.

CHAPTER ELEVEN

CASE STUDIES

Detailed Analysis of Successful Bio-Products

In this chapter, we delve into a selection of case studies that offer insights into the journeys of successful bio-products. These stories demonstrate the diverse applications and impacts of bio-products across various industries, shedding light on the challenges faced and lessons learned along the way.

One noteworthy case study is the rise of bio-based plastics, exemplified by companies like NatureWorks. NatureWorks developed Ingeo, a bio-based polymer derived from plant sugars, primarily corn starch. Ingeo offers a sustainable alternative to petroleum-based plastics, reducing greenhouse gas emissions and decreasing reliance on fossil fuels. Through innovation and collaboration, NatureWorks has overcome technical challenges to create bioplastics suitable for a wide range of applications, from food packaging to textiles.

Another compelling case study involves biofuels, with a focus on the success of the Brazilian sugarcane ethanol industry. Brazil has harnessed sugarcane's high-energy content to produce ethanol, a renewable biofuel that powers a significant portion of the country's vehicles. The success of Brazilian sugarcane ethanol demonstrates the potential of biofuels to reduce greenhouse gas emissions and decrease dependence on imported fossil fuels.

The pharmaceutical industry presents a fascinating case study with the development and commercialization of genetically engineered insulin. Companies like Genentech and Eli Lilly have played pivotal roles in revolutionizing insulin production through recombinant DNA technology. Genetically engineered insulin, which is nearly identical to human insulin, has improved the lives of millions of people with diabetes and reduced the reliance on animal-derived insulin.

In the realm of agriculture, the case study of Bt cotton highlights the impact of biotechnology on crop protection. Bt cotton incorporates genes from the bacterium Bacillus thuringiensis, enabling the plant to produce a protein toxic to specific insect pests. This innovation has reduced the need for chemical pesticides, benefiting both farmers and the environment. However, challenges related to resistance management and intellectual property rights have emerged, underscoring the complex nature of biotech crop adoption.

A noteworthy case study in the realm of consumer products is Method, a company that has successfully integrated sustainable and bio-based ingredients into its cleaning products. Method's commitment to transparency, eco-friendly formulations, and stylish packaging has resonated with environmentally conscious consumers. This case illustrates the power of branding, consumer education, and innovative marketing in promoting bio-products.

The journey of Tesla in the automotive industry provides a compelling case study in the energy sector. Tesla's electric vehicles (EVs) represent a significant shift away from traditional gasoline-powered cars. By harnessing clean and renewable energy sources, such as lithium-ion batteries and solar power, Tesla has disrupted the automotive industry, accelerating the transition to sustainable transportation.

Lessons Learned

The case studies of successful bio-products not only offer inspiring stories of innovation but also valuable lessons for individuals, businesses, and policymakers. These lessons underscore the potential and challenges of bio-products and can guide future endeavors in sustainable and eco-friendly solutions.

Innovation and Collaboration

One of the recurring themes in these case studies is the power of innovation and collaboration. Successful bio-products often result from groundbreaking advancements in science and technology. Whether it's the development of genetically engineered insulin or bio-based plastics, these innovations require cross-disciplinary expertise and collaboration between researchers, engineers, and industry professionals. The lesson here is clear: fostering an environment of innovation and collaboration can drive the creation of bio-products with the potential to transform industries.

Market Demand and Consumer Engagement

Consumer demand plays a pivotal role in the success of bio-products. Companies like Method and Tesla have demonstrated that there is a significant market for sustainable and eco-friendly alternatives. The lesson here is that understanding consumer values, engaging with them transparently, and aligning products with their preferences can lead to market success. Educating consumers about the environmental and ethical benefits of bio-products is crucial in fostering demand.

Regulatory Framework and Intellectual Property

Several case studies, particularly those in biotechnology and agriculture, highlight the importance of regulatory frameworks and intellectual property considerations. Genetically engineered insulin and Bt cotton faced regulatory hurdles and patent disputes. The lesson is that navigating the regulatory landscape and managing intellectual property rights are critical aspects of bringing bio-products to market. Effective regulation can ensure safety and compliance, while clear intellectual property rules can incentivize innovation.

Sustainability Across the Value Chain

Sustainability considerations extend beyond product development. Companies like NatureWorks have shown that sustainability must be integrated across the entire value chain. From sourcing renewable feedstocks to optimizing production processes and ensuring responsible disposal, bio-product sustainability requires a holistic approach. The lesson here is that sustainability should not be an afterthought but a fundamental aspect of bio-product development and production.

Education and Transparency

In the consumer product sector, Method's success underscores the importance of education and transparency. Consumers are more likely to embrace bio-products when they understand their benefits and trust the companies behind them. The lesson is that businesses should invest in consumer education, transparent labeling, and communication of their eco-friendly practices and ingredient choices.

Balancing Innovation and Responsibility

Tesla's case study in the automotive industry highlights the importance of balancing innovation with responsibility. While pursuing revolutionary electric vehicles, Tesla also prioritized safety, environmental sustainability, and social responsibility. The lesson is that businesses should aim for both technological advancement and ethical responsibility, as these elements can coexist and enhance brand reputation.

Continuous Improvement and Adaptation

The bio-products discussed in these case studies have not rested on their laurels but have embraced continuous improvement and adaptation. Innovations in biofuel production, for example, continue to drive efficiency and reduce environmental impact. The lesson is that success in the bio-product space often depends on a commitment to ongoing research, development, and refinement.

Consideration of Ethical and Environmental Impact

In all these case studies, ethical and environmental impact is a critical consideration. Bio-products have the potential to address societal and environmental challenges, but they must do so responsibly. The lesson here is that bio-products should aim to maximize positive impacts while minimizing negative ones, taking into account factors like biodiversity protection, fair labor practices, and carbon emissions reduction.

CHAPTER TWELVE

YOUR ROLE IN THE BIO-PRODUCT REVOLUTION

How Consumers and Innovators Can Contribute

The bio-product revolution is not limited to the realm of industry and technology; it invites active participation from both consumers and innovators. Here, we explore the roles that individuals and organizations can play in driving forward this transformative movement towards more sustainable and eco-friendly products.

Consumers as Change Agents

Consumers wield tremendous influence in the bio-product revolution. By making conscious choices, they can drive demand for sustainable alternatives. One of the most impactful ways consumers can contribute is by prioritizing bio-products in their purchasing decisions. Choosing biodegradable packaging, organic foods, and bio-based materials sends a clear signal to manufacturers that sustainability matters. Additionally, consumers can support companies that are transparent about their eco-friendly practices and ethical values, further incentivizing responsible production.

Education and Awareness

Consumer engagement extends beyond purchasing power. Educating oneself and others about bio-products and their benefits is a critical contribution. Sharing information on social media, participating in eco-conscious communities, and advocating for sustainable practices can raise awareness and inspire more individuals to embrace bio-products. An informed consumer base can drive market change and encourage businesses to prioritize environmental responsibility.

Demand for Ethical and Transparent Practices

Consumers have the power to demand ethical and transparent practices from businesses. They can inquire about supply chain transparency, fair labor standards, and sustainable sourcing. By supporting companies that adhere to these principles and holding others accountable, consumers can promote responsible production throughout the industry. Ethical consumerism is a driving force behind the bio-product revolution.

Innovators Leading the Way

Innovators and entrepreneurs are at the forefront of the bio-product revolution. Their creativity and dedication are essential for developing new, sustainable solutions. Innovators can contribute by actively seeking out opportunities to create bio-based alternatives in various industries, from agriculture and manufacturing to healthcare and energy. Collaborating with scientists, researchers, and other experts can accelerate innovation and bring bio-products to market.

Research and Development

Investing in research and development (R&D) is fundamental to bio-product innovation. Innovators can prioritize R&D efforts to refine existing bio-products, develop novel solutions, and overcome technical challenges. By continually pushing the boundaries of biotechnology and sustainable practices, innovators can expand the scope and impact of bio-products.

Collaboration and Networking

Collaboration is a driving force in the bio-product revolution. Innovators can engage in partnerships and networks to share knowledge, resources, and expertise. Collaborative efforts can help overcome obstacles and leverage collective intelligence to advance the development and adoption of bio-products. Joining industry associations, research consortia, and innovation ecosystems can facilitate these valuable connections.

Advocacy for Supportive Policies

Innovators can play a role in advocating for policies that support bio-products. Engaging with policymakers and advocating for regulatory frameworks that encourage sustainable practices, research incentives, and investment in bio-product development can create an environment conducive to innovation. Policy advocacy can help remove barriers and facilitate the growth of the bio-product sector.

Balancing Innovation with Responsibility

Innovators should prioritize responsible innovation. While pushing boundaries is important, ethical and environmental considerations should guide the development process. Responsible innovation encompasses considerations like biodiversity protection, ethical sourcing, and waste reduction. Innovators have the opportunity to demonstrate that profitability and sustainability can coexist.

Resources and Getting Involved

Getting actively involved in the bio-product revolution requires access to resources and engagement with initiatives that promote sustainability and eco-friendly solutions. In this chapter, we explore the resources available to individuals and organizations, along with avenues for getting actively engaged in the transformative movement towards bio-products.

Educational Resources

A key starting point for anyone interested in the bio-product revolution is education. Numerous online courses, webinars, and educational platforms offer valuable insights into biotechnology, sustainability, and eco-friendly practices. Universities and institutions also provide academic programs in bio-based industries, equipping individuals with the knowledge and skills to contribute effectively.

Research and Innovation Hubs

Research and innovation hubs are centers of excellence where individuals and organizations can collaborate, access cutting-edge research, and engage in bio-product development. These hubs often have state-of-the-art laboratories and experts who can provide guidance and mentorship. Joining or partnering with such hubs can be instrumental in advancing bio-product initiatives.

Business Incubators and Accelerators

Entrepreneurs and innovators looking to develop bio-products can benefit from business incubators and accelerators. These programs offer mentorship, funding opportunities, and networking support. They can help transform innovative ideas into viable bio-product businesses, facilitating market entry and growth.

Industry Associations and Networks

Industry associations and networks bring together professionals and organizations with a shared interest in bio-products. Joining such associations provides access to industry insights, networking events, and collaborative opportunities. These platforms are ideal for staying informed about industry trends and connecting with like-minded individuals.

Funding and Grants

Financial support is often crucial for research and development in the bio-product sector. Governments, foundations, and private organizations offer grants, funding programs, and competitions to support innovative bio-product projects. Securing funding can be a critical step in turning ideas into tangible solutions.

Public Engagement Initiatives

Many organizations and initiatives focus on public engagement and education related to bio-products and sustainability. Engaging with these initiatives can involve volunteering, participating in awareness campaigns, or attending events and workshops. These activities help spread the word about the benefits of bio-products and sustainable practices.

Advocacy and Policy Efforts

Advocacy plays a vital role in shaping policies that support bio-products and sustainability. Individuals and organizations can engage in advocacy efforts by joining advocacy groups, participating in lobbying activities, and working with policymakers to promote regulations conducive to bio-product development.

Collaboration and Partnerships

Collaboration is at the heart of the bio-product revolution. Building partnerships with researchers, businesses, and other stakeholders can accelerate bio-product innovation. Collaborative projects and initiatives often receive broader support and resources, leading to more impactful outcomes.

Consumer Choices

Consumers play a significant role in driving demand for bio-products. By consciously choosing eco-friendly alternatives in their everyday purchases, individuals can influence market trends and encourage businesses to prioritize sustainability. Supporting ethical and transparent companies sends a powerful message.

Knowledge Sharing and Awareness

Sharing knowledge and raising awareness are essential contributions to the bio-product revolution. Individuals can use social media, blogs, and other platforms to disseminate information about bio-products, sustainability, and responsible consumption. Educating peers and communities can have a cascading effect, inspiring others to get involved.

Conclusion:

Getting involved in the bio-product revolution is not limited to one path or approach. It is a collective effort that encompasses education, collaboration, advocacy, and responsible consumer choices. By leveraging available resources and actively engaging with initiatives and organizations, individuals and organizations can play an integral role in driving forward the transformative movement towards sustainable and eco-friendly bio-products. Together, we can shape a more responsible and environmentally conscious future.

CONCLUSION

Recapitulation of the evolution of bio-products

In this journey through the pages of "Origins of Bio-Products: The Bio-Industry Backstory," we have embarked on a captivating exploration of the evolution of bio-products and their profound impact on our world. We have traversed the historical landscapes, delved into the minds of visionaries, and witnessed the rise of sustainable and eco-friendly solutions that have the power to redefine our industries and lifestyles. As we recapitulate this remarkable journey, we are reminded of the pivotal role bio-products have played in addressing environmental challenges, fostering ethical practices, and reshaping the future.

Looking ahead, the future landscape is both promising and challenging. Bio-products, with their potential to revolutionize global industries, stand at the forefront of sustainable innovation. The lessons learned from successful case studies, the contributions of consumers and innovators, and the collaborative spirit of the bio-product revolution are powerful forces propelling us towards a more responsible and environmentally conscious world.

However, we also face significant hurdles. The bio-product revolution requires continued dedication to research, responsible practices, and ethical considerations. Striking a balance between innovation and responsibility is paramount. It is our collective responsibility to ensure that bio-products are not only environmentally sustainable but also socially and economically beneficial.

As we conclude this journey, let us remain vigilant in our commitment to sustainable and eco-friendly solutions. The evolution of bio-products is ongoing, and the future landscape holds boundless possibilities. With shared determination, we can shape a future where bio-products are at the heart of a more ethical, sustainable, and environmentally conscious world. Together, we are the architects of this transformative future, where bio-products lead the way towards a brighter and more responsible tomorrow.

Final thoughts on the future landscape

In the bio-product revolution, we witness a transformative force that transcends industries and boundaries. Bio-products offer sustainable alternatives that address pressing environmental challenges, while also fostering ethical and responsible practices. This journey has unveiled the potential of bio-products to redefine how we approach agriculture, manufacturing, healthcare, and energy, paving the way for a more responsible and eco-conscious world.

Yet, as we embrace this promising future, we are reminded of the need for balance. The bio-product revolution is not without its challenges, from responsible innovation to ethical considerations and responsible consumption. We must navigate this path with a steadfast commitment to sustainability and responsible practices, ensuring that the benefits of bio-products are maximized while their potential drawbacks are minimized.

In the end, the future landscape is one where bio-products are at the forefront of a more responsible and environmentally conscious world. It is a landscape where individuals, innovators, and consumers join forces to shape a brighter tomorrow, where innovation and responsibility coexist, and where the bio-product revolution continues to inspire and transform our industries and lives.

APPENDIX

References for Further Reading

As you delve deeper into the world of bio-products and the bio-industry, you may find the following references to be valuable sources of additional information and insights:

1. Bio-Based Materials and Biotechnology by J. M. Weydert and A. Meister. This comprehensive book provides an in-depth overview of bio-based materials, their production, and applications.

2. Sustainable Biotechnology: Sources of Renewable Energy by Om V. Singh and D. P. Singh. Explore the role of biotechnology in the production of sustainable and renewable energy sources.

3. Biotechnology for Beginners by Reinhard Renneberg. This accessible introduction to biotechnology covers key concepts and applications in an easy-to-understand manner.

4. Sustainable Agriculture and Food Security in an Era of Oil Scarcity: Lessons from Cuba by Julia Wright. Gain insights into sustainable agricultural practices and food security from Cuba's unique experiences.

5. Green Chemistry and Sustainable Technology: Challeng-

es and Prospects by R. A. Sheldon. This book explores the principles of green chemistry and their application in sustainable technology.

6. Bioproducts From Canada's Forests: New Partnerships in the Bioeconomy by A. R. Gillespie and S. L. N. Tessier. Learn about the bioeconomy and the role of Canada's forests in bio-product development.

7. The Bioeconomy to 2030: Designing a Policy Agenda by Organization for Economic Co-operation and Development (OECD). Delve into the policy considerations and strategies for fostering a bio-based economy.

8. Bio-Based Polymers and Composites edited by R. A. Gross and B. Kalra. This book explores the development and applications of bio-based polymers and composites.

9. Sustainable Energy - Without the Hot Air by David J.C. MacKay. Gain a comprehensive understanding of sustainable energy options and their potential impact on our future.

10. Bio-Based Chemicals: Value-Added Products from Biorefineries edited by A. Pandey, S. Negi, and C. R. Soccol. Explore the production of bio-based chemicals and their role in sustainable industries.

Glossary of Terms

This glossary provides definitions for key terms and concepts used throughout the book, "Origins of Bio-Products: The Bio-Industry Backstory."

1. Bio-Products: Products derived from renewable biological resources, such as crops, plants, or microorganisms, often used as sustainable alternatives to traditional, petroleum-based products.

2. Biotechnology: The use of living organisms, biological systems, or their derivatives to develop or modify products and processes for specific purposes, such as medicine, agriculture, or industry.

3. Biodegradable: Capable of being broken down naturally into harmless substances by microorganisms or other biological processes, reducing environmental impact.

4. Circular Economy: An economic model that promotes resource efficiency and sustainability by reducing waste, reusing products, and recycling materials in a closed-loop system.

5. Genetic Engineering: The manipulation of an organism's genetic material to introduce desirable traits or characteristics, often used in the production of genetically modified organisms (GMOs) and biopharmaceuticals.

6. Sustainability: The practice of using resources in a way that meets current needs without compromising the ability of future generations to meet their needs, encompassing environmental, social, and economic considerations.

7. Ethical Considerations: Moral principles and values that guide decisions and actions, especially in relation to responsible and sustainable practices in bio-product development.

8. Renewable Energy: Energy derived from sources that are naturally replenished, such as solar, wind, and biomass, as opposed to finite fossil fuels.

9. Transparency: Openness and clarity in business practices, including the disclosure of information about product sourcing, production processes, and environmental impact.

10. Consumer Consciousness: Awareness among consumers about the environmental, ethical, and social implications of their purchasing choices, leading to more responsible consumption.

11. Biodegradable Plastics: Plastics designed to break down naturally into environmentally benign substances, reducing plastic waste.

12. Carbon Footprint: The total greenhouse gas emissions, primarily carbon dioxide (CO_2), associated with a product, process, or individual's activities, often used to assess environmental impact.

13. Recycling: The process of collecting, processing, and reusing materials to create new products, reducing waste and conserving resources.

14. Green Chemistry: The design of chemical products and processes that minimize the use and generation of hazardous substances, promoting environmental sustainability.

15. Circular Economy: An economic model that promotes resource efficiency and sustainability by reducing waste, reusing products, and recycling materials in a closed-loop system.

ACKNOWLEDGMENTS

The creation of "Origins of Bio-Products: The Bio-Industry Backstory" has been a journey enriched by the contributions and support of numerous individuals and organizations. We extend our heartfelt gratitude to those who have played a pivotal role in bringing this book to fruition.

First and foremost, we express our appreciation to the researchers, scientists, and experts whose groundbreaking work in the field of bio-products provided the foundation for this exploration. Your dedication to sustainable and eco-friendly solutions is truly inspiring.

We are indebted to the educators and institutions that have nurtured a culture of learning and inquiry, fostering a deeper understanding of biotechnology, sustainability, and responsible practices.

To the innovative companies and entrepreneurs who are at the forefront of the bio-product revolution, we thank you for your vision, commitment, and willingness to share your experiences. Your case studies and insights have enriched the narrative.

To our readers, your curiosity and interest in the bio-product revolution have driven us to deliver a comprehensive and informative account. We hope this book serves as a valuable resource on your journey toward a more sustainable future.

Finally, to our friends and families, whose unwavering support and encouragement sustained us throughout this endeavor, we offer our deepest gratitude.

The evolution of bio-products and the bio-industry is a collective effort, and we are humbled to have been a part of this transformative journey. Thank you for joining us in exploring the origins and potential of bio-products in shaping a more responsible and environmentally conscious world.